READ ALL ABOUT IT!

GENETICS

ROBERT SNEDDEN

W
FRANKLIN WATTS
LONDON·SYDNEY

This edition 2004

First published in 2000 by Franklin Watts
96 Leonard Street, LONDON EC2A 4XD

Franklin Watts Australia
45-51 Huntley Street
NSW 2015

Copyright © Franklin Watts 2000 and 2004

Series editor: Rachel Cooke
Assistant editor: Kate Newport
Designer: John Christopher, White Design
Picture research: Sue Mennell

A CIP catalogue record for this book is available from the British Library.

ISBN 0 7496 5675 1

Dewey Classification 174

Printed in Malaysia

Acknowledgements:
Cartoons and artwork: Andy Hammond pp 6, 9, 22, 27; Sholto Walker p. 17; Mark Watkinson pp 8, 15.
Photographs: Front cover: Ronald Grant Archive: centre right below; Oxford Scientific Films: top right (Richard Kolar/Earth Sciences), bottom right (Scott Camazine); Popperfoto: main picture (Reuters), centre right above (Reuters);
Back cover: Popperfoto (Reuters)
Insides: Associated Press AP p. 9 (Judi Bottoni);
Corbis p. 10b (AFP); Format Photographers pp. 5b (Maggie Murray), 20 (Jenny Matthews), 28b (Ulrike Preuss); Frank Lane Picture Agency p. 3 tl (B.B. Casals); Panos Pictures p. 10t (Sean Sprague); Popperfoto pp. 3 tc (Reuters), 3tr (Romeo Gacad/AFP), 7 (Jeff Mitchell/ Reuters), 11 (Reuters), 18 (Steve Dipaola/Reuters), 19 (Reuters), 22 (Dan Chung/Reuters), 28t, 29 (Romeo Gacad/AFP); Oxford Scientific Films pp. 6 (Keith Ringland), 8 (Richard Kolar/Earth Sciences), 19b (Reuters); Ronald Grant Archive pp. 16, 25t (Columbia Tristar Films (UK)), 25b (UIP); Science Photo Library pp. 3b (Eurelios/Plailly), 4 (Mauro Fermariello), 5t (James King-Holmes), 12 (A. Barrington Brown), 14t (Hank Morgan), 14b (David Parker), 15 (Simon Fraser), 19t (Gary Parker), 21 (Philippe Plailly), 23 (Alfred Pasieka), 24 (Andrzej Dudzinski), 26t (Tom McHugh/Field Museum, Chicago), 26b (BSIP/Boucharlat), 27 (Philippe Plailly).

EDITOR'S NOTE

Read All About It: Genetics takes the form of a newspaper called *The Genetics News*. In it you can find articles about a lot of different subjects and many facts. It also includes opinions about these facts, sometimes obviously as in the editorial pages, but sometimes more subtly in a news article: for example in the article concerning pig clones (page 19). Like any newspaper, you must ask yourself when you read the book 'What does the writer think?' and 'What does the writer want me to think?', as well as 'What do I think?'.

However, there are several ways in which *The Genetics News* is not and cannot be a newspaper. It deals with one issue rather than many and it has not been published on a particular day at a particular moment in history, with another version to be published tomorrow. While *The Genetics News* aims to look at the major issues concerning genetics and its associated technologies, the events reported have not necessarily taken place in the past few days but rather over the past few years. They have been included because they raise questions that are relevant to the issue today and that will continue to be so in the future.

Another important difference is that *The Genetics News* has been written by one person, not many, in collaboration with an editor. He has used different 'voices' but the people and events reported and commented on are real. In addition, the letters on page 17 are extracts from other people's writing on the issues surrounding genetics and are credited accordingly.

There are plenty of other things in *The Genetics News* that are different from a true newspaper. Perhaps a useful exercise would be to look at the book alongside a real newspaper and think about, not only where we have got the approach right, but where we have got it wrong! Enjoy reading *The Genetics News*.

THE GENETICS NEWS

| Home News 7 — GM pollen in honey  | Foreign News 11 — Europeans' GM fears  | Sport 28 — Genetically enhanced?  |

GENOMANIA

The secrets of the human genome are revealed

2003 will go down in history as the year when the detailed map of the human genome – the 30,000 or so genes that make up the recipe for a human being - is completed.

The News Editor

Having the complete genetic blueprint for humans will transform medicine. New drugs, new treatments and new definitions of disease, based not on symptoms but on genetics, will unfold before us. With knowledge of your unique genetic profile, your doctor may some day be able to diagnose diseases before their effects even show up and prescribe individually tailored medications to deal with them.

There might be other, not-so-positive consequences – the human genome will not only help doctors but it may help insurance companies decide on your insurance premiums! Doubt has also been cast on some of the motives behind the research. Is it the desire to advance human knowledge or to exploit the potentially vast profits to be made by selling that knowledge to the drug companies? ■

THE RACE TO THE FINISH: FOR THE FULL STORY SEE PAGE 14

GOODBYE DOLLY, HELLO DOTCOM

Remember Dolly the sheep? She hit the headlines back in 1997, the first mammal to be successfully cloned, an exact genetic copy of her 'mother'. She may have been the first, but she certainly wasn't the last. Today Dolly is no more, her remains preserved in a glass case in Edinburgh, but the cloning goes on. Cows, horses, monkeys, mice and pigs have all been cloned now. On page 19 you can find out about Dotcom the pig and her sisters. ■

Dolly – a clone at home on the range!

INSIDE: Home news 4–7, Foreign news 8–11, Genetic Profiles 12–15, Editorial and Letters 16–17, Cloning news 18–19, Health 20–21, Business and Farming 22–23, The Culture 24–25, Origins 26, Gardening 27, Sport 28–29, Directory 30–32

Health costs to rocket

The Government have issued a White Paper, 'Our inheritance, our future - realising the potential of genetics in the NHS'. According to the Department of Health, it sets out 'a vision of how patients could benefit in future from advances in genetics'. It is planned to spend over £50 million on genetics in healthcare over the next three years. The Secretary of State for Health said: 'Our vision is for the NHS to lead the world in taking maximum advantage of the safe, effective and ethical application of the new genetic knowledge and technologies'. The Human Genetics Commission welcomed the White Paper.

The Government has given the HGC the job of looking into the issues involved in genetically screening all babies at birth and storing the data for future use. GeneWatch UK warned that people would not be protected from genetic discrimination and that there was a need to avoid creating a 'genetic underclass' (see story opposite 'Discrimination fears unfounded?').

STRATEGY NEEDED

Geneticists were happy that the Government were 'taking a step in the right direction' but most felt that the funding was inadequate, with some even thinking that Britain would fall behind the rest of the world, particularly in gene therapy research.

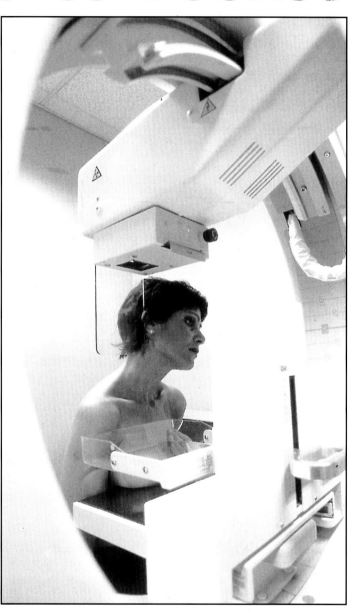

Genetic screening could target breast cancer.

THE MEANING OF SCREENING

Among the major impacts of genetics developments currently being looked at by the government and its advisers is the possible introduction of mass genetic screening programmes and the establishment of a national DNA testing network. Screening programmes to identify those people most vulnerable to diseases such as breast cancer, Alzheimer's and schizophrenia could become commonplace in the near future.

RISK ANALYSIS

At the moment, doctors diagnose symptoms and recommend treatment. In the future genetic screening will make them aware of the risk of a person developing certain illnesses and the need to take action to reduce that risk. Even where there is not a genetic component to a disease, people's genetic profiles may reveal that they are more likely to suffer from it than someone else.

Ministers are concerned that the impact of genetics on health will be even greater than the concern created over genetically modified foods. For example, there could be implications for the welfare system because genetic testing would make it possible to predict who was likely to become disabled or die too young to need a pension. Will people be happy for governments to have this kind of insight into their future health – or ill-health?

Private Perils

Genetic tests – an opportunity for private medical companies

The News Editor

Private medical firms are set to make a killing from genetics by providing tests and screening that the NHS may not be able to offer because of inadequate funding. Tests giving information on diet and lifestyle based on an individual's genetic profile are already available, with tests for genes linked to heart disease, obesity and skin complaints on the way.

Concern has been expressed about how these private tests will be regulated. The Human Genetics Commission has published a report, Genes Direct, that makes some key recommendations. They suggest that genetic testing should be regulated. The important thing is to safeguard the public from poor quality testing and ensure that the patient receives proper back-up and advice on the results of their test.

A scientist studies the results of a screening.

DISCRIMINATION FEARS UNFOUNDED?

Disabled fear further discrimination in the workplace.

A survey carried out recently in Britain by the disability charity Radar found that 90 per cent of disabled people feared that the growing emphasis on genetics would lead to further discrimination, and wanted a ban on employers and insurers having access to information suggesting someone might develop a genetic condition.

Commenting on a Government White Paper on genetics, Dr Helen Wallace of GeneWatch warned that genetic testing to identify individuals who were likely to contract certain illnesses, could lead to a 'genetic underclass'.

'The government has put the interests of industry above those of people,' said Dr Wallace. 'There is a shocking lack of safeguards for people taking genetic tests. The Paper does not address the gaping holes in the legislation.'

NO LOGICAL BASIS

No legal safeguards are proposed by the White Paper that would prevent companies from misusing any data they obtained. The emphasis on genes ignored other factors that contributed to disease such as environmental factors. Some genetic counsellors have also pointed out that knowledge of the full complement of human genes may prove that everybody has some kind of abnormality. If this were so then it would eliminate any logical basis for discrimination.

IS IT A COD? IS IT A PLAICE? NO, IT'S SUPERFISH

Home Affairs Editor

English Nature, the Government's conservation advisers, have warned that they will oppose the release of genetically modified superfish in British waters unless the fish are made infertile first. Scientists believe that the GM superfish would replace natural fish within a few generations if they were released into the wild.

Salmon, carp and tilapia are the main species being studied by the geneticists. A technique for transferring growth promoter genes from other species has led to a number of research projects that aim to make fish farming more efficient. Injecting growth promoter genes into salmon eggs can make them grow four times faster than in the wild and the transgenic (where genes have been introduced from another species) fish end up being 22 per cent larger on average than non-transgenic fish.

Because big fish are more likely to breed than little fish the number of transgenic fish in a population could rise from 1 in 100,000 to 1 in 2 within 16 generations.

GM FISH ESCAPE

In Britain all experiments on GM fish were halted after concerns were raised that rules on containment at experimental salmon tanks beside Loch Fyne, Argyll, were not strict enough. English Nature say that genetic pollution has already occurred when fish from Scottish fish farms have escaped into the wild as a result of bad weather and changed the wild salmon stocks. Some conservationists believe that farmed fish breeding with Norwegian salmon may have had an effect on the wild fish's ability to survive in small west coast rivers and that this could be a factor in the dramatic decline in salmon.

What's leaping out of the gene pool?

NEW HOPE FOR HEROIN ADDICTS

Scientists at University College London (UCL) say they have uncovered the gene that provides the pleasure drug users get from heroin and other opium-related substances.

GOOD FEELINGS

Drugs act on chemical receptors in the brain to produce a feeling of pleasure and well-being. The UCL researchers identified a gene that carries the instructions for making a protein that acts as a receptor for a chemical in the brain, called substance P. This turned out to be one of the pleasure-producing chemicals in opium-based drugs, including heroin. The scientists were able to create genetically engineered mice that do not become addicted to these opiates.

The report in *Nature* magazine said that the discovery could lead to safer pain-killing drugs and open up new ways to help drug addicts defeat their habit. Stephen Hunt from UCL said: 'The gene seems to separate the analgesic (pain-relieving) properties of opiates from their pleasurable-addictive effects.'

A taste of honey

Traces of pollen from GM oilseed rape found in honey

For some GM food has a sour taste – honey findings add further fuel to the protestors' demands.

The government's recent decision to allow the production of some genetically modified crops has raised fears amongst organic farmers and environmental groups. Tony Jupiter, Director of Friends of the Earth, said 'Tony Blair must not ignore the threat GM poses to our food and the environment.'

The go-ahead comes despite a series of cross-contamination scares concerning the outdoor testing of GM plants.

GM POLLUTION

After traces of pollen from genetically modified crops were found in honey made near test sites, Friends of the Earth claimed that GM pollen was threatening the livelihood of organic farmers and beekeepers working within 3 kilometres of the text crops. As a result, the Bee Farmers Association (UK) warned its members to move their hives at least 10 kilometres away from the nearest GM crops.

A government panel was then set up to report on these scares. It found that, although there were no known health effects, it would be almost impossible to grow some GM crops without contaminating fields of the same species.

What lies in store for organic farmers is not clear but they will be relieved that species of GM crops where cross-contamination is 'almost inevitable' will not be permitted.

EXCLUSIVE! NATIVE AMERICANS ARE RUSSIAN

Ilya Zakharov, deputy director of Moscow's Vavilov Institute of General Genetics, claims to have taken the first steps towards pinpointing the precise origin of native Americans – and the path leads to the Tuvan people in Siberia.

Dr Zakharov said that an expedition he led had uncovered a DNA link between native Americans and the Tuvan people of the Ak-Dovurak region 3400 km southeast of Moscow. Zakharov is not the first to suggest a Siberian connection for the Americans but he may be the first to produce evidence for it. 'This is a big breakthrough,' he said. 'We had examined a lot of populations before – and by pure chance the results proved it was the Tuvans.'

Dr Zakharov analysed DNA from hair samples taken from about 430 Tuvans and then compared it with that of Inuit and Amerindian people including the Navajo and Apache.

DNA MATCH

What Zakharov found was that the Amerindian DNA exactly matched the Tuvans' – by 72 per cent of one group of 30 samples and 69 per cent of another group of 300, 'the highest frequencies of Amerindian DNA types ever found,' he says. He now wants to look at the DNA of people from the Khakassia and Altai regions, which border on Tuva. He believes that these people may have an even higher DNA match with Amerindians. ■

It seems Russia conquered America after all!

WHO ARE THE TUVANS?

Tuva bridges Siberia's huge Taiga Forest and the steppe grasslands north of Mongolia. It is one of Russia's poorest and most mysterious regions, with ancient cultural traditions that include shamanism, an ancient religion based around magic and sorcery. The Tuvans are mainly nomadic pastoralists who herd camels, yaks, sheep, goats and reindeer.

Scientists have long believed that some 30,000 to 40,000 years ago people migrated from Asia across the ice sheets of Siberia's Bering Strait to Alaska. Previously geneticists speculated that America's first inhabitants, possibly no more than 5,000 people, originally came from Northern China or Mongolia. ■

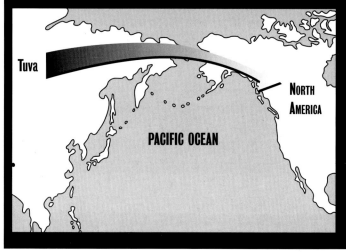

Catch up

Genetics research in Russia was held back for decades by Trofim Lysenko, a crop scientist who rejected Gregor Mendel's ideas about inheritance because they didn't fit in with Communist thinking. Lysenko was Stalin's favourite scientist and anyone who dared to disagree risked being exiled to Siberia's prison camps, or worse. Russian scientists have had to play catch up with the West ever since. ■

Gene-therapy death

Research methods questioned after lethal results

Foreign Affairs Editor

Jesse Gelsinger was an 18-year-old research volunteer for gene-therapy at the University of Pennsylvania. He suffered from a rare liver disorder, caused by a genetic mutation. His condition could be controlled by drugs and diet, and he did not think that gene therapy would cure him but he hoped it might help babies who were born with a more fatal form of the disease.

Four days after scientists infused genetically engineered viruses into his liver Jesse was dying as his blood thickened like jelly, clogging vessels, and his kidneys, liver and brain failed. So Jesse Gelsinger became gene therapy's first fatality.

Little real progress has been made in a field that was at one time thought to hold great promise. The crucial difficulty to overcome is how to get the genes safely to the target cells in the patient's body. Gelsinger developed a fatal immune response to the virus being used to deliver the genes to his cells. His death dealt a severe blow to the advancement of gene therapy.

Another setback came when French researchers reported that they had successfully cured four boys of severe combined immunodeficiency, or SCIDs, also known as the 'boy in the bubble' syndrome because sufferers have to be kept isolated from contact with any disease-causing agents.

Dr Krishna Fisher, part of the Pennsylvanian team and later a critic of its over-hasty methods of research.

Another setback

Optimism faded within a few months when it was discovered that some of the inserted genes had been delivered to the wrong places, causing two of the boys to develop leukaemia.

Despite the setbacks the enthusiasm for new experiments continues. Scientists at Imperial College, London, and the Medical Research Council are working on a technique that uses microbubbles and ultrasound instead of viruses. Tiny bubbles mixed with DNA are injected into the muscles and then popped using the ultrasound. This makes microscopic holes in the cells through which the DNA can enter. It is hoped that it could be used to treat children with muscular dystrophy. ■

The Clonesome Pine

An 800-year old pine tree in Korea is to be preserved using genetic engineering. In 1464, so legend has it, a Korean king was passing down the road where the tree stands on his way to Pobju Temple. When the king complained that the tree's branches were in the way, they immediately rose upwards to allow him to pass through.

Today, the umbrella-shaped pine, designated a national monument in 1962, is withering and the government's Institute of Forest Protection and Management Research has decided to replicate the tree by copying and cultivating its cells to preserve the genetic traits of the pine. The genetic engineering is similar to cloning animals from cells (see page 18). ■

GM CROPS TO FEED THE WORLD

Genetically modified crops could help developing nations solve their food problems, according to Geoffrey Hawtin, director-general of the International Plant Genetic Resources Institute (IPGRI).

People in developed countries could afford the luxury of rejecting such food, but for countries such as China and India the high yields and disease resistance of GM crops could bring huge benefits. Hawtin points out that the world's poorest people are spending 90 per cent of their income on food and for them 'the risk [of eating GM foods] is minute compared with the risk of having not enough food to eat.'

Sandy Thomas, director of the Nuffield Council for Bioethics, declared that, 'We do not claim that GM crops will eliminate the need for economic, political or social change, or that they will feed the world. However, we do believe that GM technology could make a useful contribution, in appropriate circumstances, to improving agriculture and the livelihood of poor farmers in developing countries.'

GM crops could transform life for poor farmers.

BIOTECH, BIOTALK

Biotech? No thanks!

Both sides claimed success after demonstrators rallied to disrupt an international conference on how genetically modified foods can help ease poverty in developing countries.

Demonstrators in Sacramento, USA, claimed to have got their message through despite a large police presence.

At the conference, the US Secretary of State for Agriculture, said 'a seed has been planted' for the advancement of bio-technology. Some say however that the US is pushing forward bio-technology to gain economic advantage without proper consideration of the risks faced by poorer nations.

A Zambian nutritionist said, 'we feel with bio-technology, that we should take our time and understand what we're dealing with. In the meantime, we would like to continue with conventional methods.'

Absent from the talks were representatives from the European Union. The United States is lobbying the World Trade Organization to force the EU to end its ban on genetically modified food.

Fear of the unknown

Government's decision conflicts with public opinion

Foreign Affairs Editor

In March 2004 the UK government gave the go-ahead for the commercial production of certain genetically modified crops. Margaret Beckett, the Environment Minister, described the decision as 'difficult, bedevilled by confusion' but 'the right one'.

This announcement conflicts with widespread public distrust of bio-technology but was welcomed by the British Medical Association, who have claimed that public attitudes are led by 'hysteria'.

Irrational or not, altering the genetic make-up of food is a politically charged issue because of perceived concerns about health and the risk to the environment. Axel Kahn, head of a European Union scientific committee on bio-technology, said, 'There is a great anxiety among Europeans on food safety.'

According to a European Commission survey many people are afraid of developments in biology because they are less informed about these issues.

Similarly, **GM Nation?** – a countrywide debate organized by the UK government, found people were 'cautious, suspicious or out-rightly hostile about GM crops'. People wanted more research and were wary of big-business involvement in GM use.

Despite this, GM food is here to stay and the pressure is now on the government to allay public fears.

A French protestor uproots the problem!

A gene-free tomato?

The survey also revealed that, when shown the statement 'Ordinary tomatoes do not contain genes, while genetically modified tomatoes do', an equal number of Europeans agreed this was true (35 per cent) as those who said it was false (35 per cent). Thirty per cent said they did not know.
Do you?

THE SPEED OF CHANGE

The US National Bioethics Advisory Commission has called for broad public discussion of the issues relating to biotechnology, which include gene patenting and the possibility of 'designer children'. The call comes as many suggest it is the pace of the biotech revolution that makes them feel uneasy. Breakthroughs come at such a furious rate that even scientists are left struggling to keep up. What chance do the ethicists, or the public, have?

With opinion already so polarised it is hard to see where this could take place but it's an issue that has to be faced. Eric Lander, director of the Whitehead Institute at the Massachusetts Institute of Technology, told his colleagues, 'There has been an explosion of issues in the past year, and we don't have the trust of the public.' Dr Lander emphasised the need to engage in thoughtful discussion, and to be seen to have that discussion 'in good faith'.

Genetic Profile 1

HE HAS HIS MOTHER'S EYES!

Crick, Watson and their model of the 'secret of life'.

Every day friends and relatives look closely at little children and claim to see a resemblance to one or other of their parents. When you think about it we all look pretty much the same really. Two eyes, a nose, two ears, usually five fingers on each hand, five toes on each foot – and so on. Apple trees grow from apple seeds, pigs give birth to pigs, chickens never hatch from duck eggs. So what is it that keeps living things looking much like their parents?

The secret of life

Characteristics are passed from parents to offspring, in a process called heredity. This seems obvious. If both your parents are tall with blue eyes then there is a good chance that you will be too. However, if one parent has brown eyes and the other has blue eyes you might have either colour yourself. You wouldn't have one brown eye and one blue eye. What wasn't obvious until about 50 years ago was how it was done.

On 28 February 1953, Francis Crick burst through the door of the Eagle pub in Cambridge and announced, 'We've discovered the secret of life.' As the project to unlock the human genome is completed it might be a good moment to take a look at the star of that momentous event – the DNA molecule.

The double helix

In nearly every cell in your body there is a codebook. This codebook isn't written in words on paper but in the structure of a giant molecule – DNA. The secret that Crick and his colleague, James Watson, had uncovered was the way DNA is put together and the way its chemical alphabet works. Watson and Crick worked out that DNA is shaped like a twisted ladder – a double helix shape. Each long strand of the ladder is made up of a huge number of units called nucleotides. These in turn are made up of units called bases.

The four bases that make up the DNA molecule are adenine (A), thymine (T), guanine (G), and cytosine (C). The bases form interlocking pairs that can fit together in only one way:

A always pairs with T;
G always pairs with C.

The pairs of bases linking up between the long strands form the 'rungs' of the ladder. Every time a cell divides, as happens in the process of growth, the DNA helix 'unzips' and each side is used as a template to form a new strand. In this way two identical double-strands of DNA are produced (see diagram above right). This

THE GENETICS NEWS — Genetic Profile 1

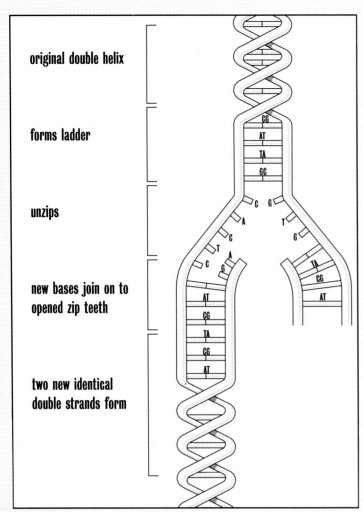

- original double helix
- forms ladder
- unzips
- new bases join on to opened zip teeth
- two new identical double strands form

160 billion kilometres of DNA

Each of the 100 trillion cells in your body (with the exception of the red blood cells) contains all the genetic information necessary to build a human being. Half of this information comes from your father and half from your mother, which is why some of your genes give you characteristics rather like your father's and some like your mothers. Inside the nucleus of each cell 2 metres of DNA is divided between 23 pairs of chromosomes – one set from each parent. There is enough DNA in your cells to stretch 160 billion kilometres!

Every living organism on earth, from microbes living in super-heated sulphur vents on the bottom of the ocean to chimpanzees (whose DNA is 98.9 per cent the same as ours), uses the same four-letter genetic code contained in DNA. We're all reading from the same book – and it is this that allows genetic engineers to perform their cut-and-paste manipulation of life. ■

enables the codebook to be copied again and again, with scarcely any mistakes, so each cell has its own copy.

Decoding the secret

The bases are the 'letters' of the DNA alphabet. Almost all of the codebook – 97 per cent, in fact – is 'junk' DNA, meaningless and unreadable. What remains is a set of instructions for making proteins. These protein recipes are spelled out using the bases to form three-letter 'words' called codons. Each codon corresponds to a particular amino acid, the chemical building blocks from which proteins are constructed. The cell uses molecules called RNA to 'read' the codons and assemble the protein according to the instructions. The recipe for a single protein is a gene.

Protein performance

Proteins are essential components of all your body's organs, they are part of your defence against disease and they also regulate all the chemical activities that go on in the body. The function of a protein depends on its shape and this is determined by the order in which its amino acids are joined together.

Proteins perform vital tasks in living organisms, regulating chemical reactions, fighting infection and providing much of the structure of larger organisms. Genes make proteins and proteins control you. Genes make you male or female and help decide your size, build, eye colour, hair colour and other features.

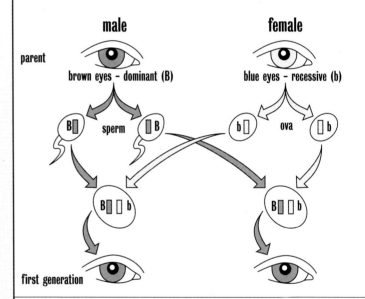

Should two brown-eyed people, both with a blue-eyed parent, have children together then this pattern would be likely.

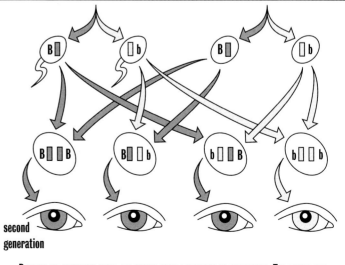

DIAGRAM TO SHOW HOW GENES ARE PASSED FROM GENERATION TO GENERATION. THE BROWN-EYED GENE (B) IS 'DOMINANT' BUT THE BLUE-EYED GENE (b) WILL STILL BE PASSED ON.

Genetic Profile 2

The race to the finish

Craig Venter

In April 2003 the International Human Genome Sequencing Consortium announced that the Human Genome Project had been successfully completed. By luck or good judgement the announcement coincided with the fiftieth anniversary of the publication of James Watson and Francis Crick's discovery of the structure of DNA.

James Watson, first director of the Human Genome Project, said: 'Never would I have dreamed in 1953 that my scientific life would encompass the path from DNA's double helix to the 3 billion steps of the human genome. But when the opportunity arose to sequence the human genome, I knew it was something that could be done – and that must be done.'

The completion of the first draft of the human genome was announced in June 2000. Racing towards that important milestone against the Consortium was the privately funded Celera Genomics of Rockville, Maryland, led by its maverick president, Craig Venter.

The competition to complete the first draft was often fierce, with Celera and the Consortium calling each other's methods into question. With in a few months of the first draft tie Celera had dropped out of the gene sequencing race and turned towards drug research instead. The field was left open to the Consortium researchers to turn the draft copy into the finished article.

The Human Genome Project now has an 99.99 percent accurate map of about 99 per cent of the human genome's gene-containing regions. The very few remaining gaps (there are about 400, compared to the 150,000 or so in the draft version) correspond to regions in the genome with unusual structures that

Using computer processing power to sequence the human genome.

THE GENETICS NEWS ■ *Genetic Profile 2*

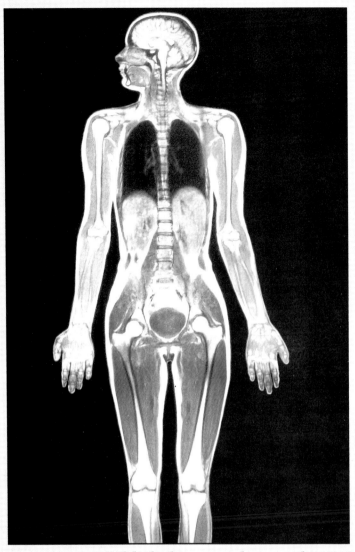

Here's one 'map' of the body – now we have another one!

cannot be reliably mapped using current technology. These regions appear to contain very few genes, however.

The International Human Genome Sequencing Consortium included hundreds of scientists at 20 sequencing centres in China, France, Germany, Great Britain, Japan and the United States. In the spirit of scientific cooperation that has guided the whole enterprise all data generated by the Consortium is freely available to scientists around the world, with no restrictions whatsoever on its use.

FREELY AVAILABLE

Genetics researchers have not been slow to make use of this treasure trove. When the Human Genome Project first began in 1990, scientists had identified fewer than 100 genes responsible for human disease. Having access to sequenced DNA can make a major difference to researchers looking for the regions of the human genome that may contain the small and often rare DNA variations that are involved in disease. Today, more than 1400 disease-causing genes variations have been discovered.

Dr Francis Collins, director of the National Human Genome Research Institute said, 'The completion of the Human Genome Project should not be viewed as an end in itself. Rather, it marks the start of an exciting new era – the era of the genome in medicine and health. We firmly believe the best is yet to come, and we urge all scientists and people around the globe to join us in turning this vision into reality.' ■

THE PATH TO THE GENOME

■ **1856** Austrian monk and botanist Gregor Mendel begins experiments in breeding peas that will lead him to the laws of heredity.

■ **1869** Swiss biochemist Johann Miescher discovers a material in cells that is now known as the genetic material DNA.

■ **1902** American geneticist Walter Sutton and German zoologist Theodor Boveri show that cell division is connected with heredity.

■ **1906** English biologist William Bateson introduces the term 'genetics'.

■ **1944** American bacteriologist Oswald Avery, American biologist Colin MacLeod, and American biologist Maclyn McCarthy demonstrate the role of DNA in genetic inheritance.

■ **1953** American biochemist James Watson and English biochemist Francis Crick discover the structure of DNA.

■ **1957** The 'central dogma' of molecular biology is established by Crick – DNA instructs RNA to make proteins.

■ **1966** The genetic code is cracked – three bases make a codon – the code for an amino acid.

■ **1967** American scientist Charles Caskey and associates demonstrate that the genetic code is common to all life forms.

■ **1973** American biochemists Stanley Cohen and Herbert Boyer develop the first techniques of genetic engineering.

■ **1981** The US Food and Drug Administration grants permission to Eli Lilley and Co to market insulin produced by bacteria, the first genetically engineered product to go on sale.

■ **1983** Scientists begin to discover the genes behind inherited diseases.

■ **1987** The first genetically altered bacteria are released into the environment in the USA; they protect crops against frost.

■ **1990** The Human Genome Project is launched.

■ **1998** The first whole animal genome is completed, that of a worm.

■ **1999** The first human chromosome is sequenced.

■ **2000** The first draft of the human genome is completed. ■

Editorial

THE FEAR OF FRANKENSTEIN

Mention the subject of genetics and many people will react with hostility. This is interfering with nature, some might say. It is something that we should not do. Mary Shelley's novel *Frankenstein*, about a scientist whose attempt to create life had tragic consequences, might be mentioned. 'Frankenstein foods' was one description the news media used to describe genetically modified foods, for example.

We are living through a time of great discovery about human genetics. It has become part of our popular culture. Like most other things, perhaps, genetics has come to be seen as either 'good' or 'bad'. Good genetics improves our ability to diagnose and treat serious diseases, such as muscular dystrophy, bad genetics could be used to create a genetically superior 'super-race', discriminate against those with 'faulty' genes or poison the environment with alien organisms.

What is life anyway?

What is 'life' anyway? Is creating life in the laboratory something we should steer clear of? A distinguished panel of biomedical ethicists, theologians and scientists came up with answers that amounted to 'no problem'.

The ethical debate was prompted by a series of experiments at the Maryland-based Institute for Genomic Research (TIGR). These experiments were carried out on the bacterium Mycoplasma genitalium, which has the least genetic material of any known organism, just 480 genes. Humans, by contrast, have around 100,000 genes. It actually appeared that the bacterium only needed about 300 of these genes to survive. To find out if this were truly all that was needed for life, the scientists would have to manufacture the genes, insert them into an engineered cell, and see whether it sprang to life.

Arthur Caplan, a world-renowned bioethicist at the University of Pennsylvania, assembled a panel of experts to debate the issues. The outcome was surprising, given current public fears about genetic modification. 'No one had any objection to minimal genome or synthetic life research,' said Caplan.

Playing God?

One issue the panel wrestled with was whether creating synthetic life amounted to 'playing God'. They decided that it didn't. They decided that intervening in the natural world is no bad thing, as long as it is done for human good. They pointed out that some of the benefits that could come from this research might include 'designer' bacteria to eat pollution or mass produce new drugs, as well as providing great scientific insights into the nature of life.

Scientific knowledge gives us a way of understanding how the world works – and we are as much a part of the world as anything else. A great deal of who and what we are is influenced by our genes. We can make choices about the things we want for ourselves and our children, but they should be choices that are informed by knowledge and understanding, not ignorance and fear. People have reasonable fears, but they are fears to be addressed by society not genetics. Should genetics be closed to scientific investigation? No, it should not. Taboos, to quote a popular science magazine, are simply a bad excuse for not thinking. Knowledge is surely better than ignorance. ■

Letters to the editor

Changing our genes
Before we start doing germ-line gene therapy [re-programming the genes in an egg or sperm cell] we need to decide whether we want to change what it is to be human. We need to decide whether there is something about human nature that is so valuable that we shouldn't change it, even if it could be done. This is involuntary intrusion into future generations. The whole thing is very chancy. There are so many ways in which we can go wrong. Who knows what the effects of our actions today will be in 2200?
Cynthia Cohen, Kennedy Institute of Ethics at Georgetown University, Washington, D.C.

Moving forward
We are gaining the power to intervene in a realm that we've never had access to... by introducing new genes into the flow from one generation to another. Some people say you can't do that, you shouldn't do that. But I think we just have to take responsibility and move forwards.
Gregory Stock, University of California, Los Angeles

Duplicate babies — no thanks
Every child should be wanted for itself, as an individual. In making a copy of oneself, a parent is deliberately specifying the way he or she wishes that child to develop... I suggest that there is a greater need to consider the interests of the child and to reject these proposed uses of cloning.
Dr Ian Wilmut, Roslin Institute, leaders in cloning research (see page 19)

Tomato source
If you saw the tomato in the wild before it was domesticated, you would see a little tiny green bitter fruit that nobody would eat. Across the last few thousand years, humanity has bred it selectively, changed its genes by selection, to give us the thing we call a tomato. Where scientists have used genetic engineering to alter one gene that's involved in ripening, that is infinitely less than all of the things that have been done to that plant in the last 2000 years. It just doesn't make any sense to be concerned about that as a foodstuff you eat... There are certainly legitimate concerns that I think everybody has about genetically modifying organisms and putting them out in the environment. The concern is that appropriate tests [be] done to make sure that they don't have unwanted effects.
Bruce Baker, professor of biological sciences, Stanford University

The voice of reason?
We shouldn't underestimate the public. People have a lot of common sense. I can see suspicion of the new and the untested. When big government and big business tell us something, [it] hasn't necessarily been to our benefit. So the public is being wary. We're doing a great experiment now in a lot of ways, genetically engineering things into the environment. It's not different in quality from breeding by conventional mechanisms, but it is a different pace, and so I think for people to question it is reasonable. Not that I'm saying we should stop, I'm just saying that there is this balance to achieve, [and it] will emerge with time.
Elizabeth Blackburn, professor and chair-woman of the Department of Microbiology and Immunology at the University of California, San Francisco

Cloning News 18

Multiplicity

Monkeys multiply with new cloning techniques

The Science Editor

Researchers in Oregon grew a live monkey from one-quarter of a monkey embryo recently, opening the possibility of creating up to four identical monkey clones from a single embryo. They have called the monkey Tetra.

The Oregon team's technique for splitting embryos had been developed as an aid to medical research. The goal was to eliminate the genetic variability of monkeys that had long been a problem for scientists studying human diseases in primates. Having animals with identical genes will allow scientists to begin to unravel the different roles played by genes and environment in the development of disease.

SIMPLER METHOD

Previous efforts to clone monkeys using the more complicated method used to produce Dolly the sheep and Dotcom the pig (see opposite) resulted in just two births in 1997 and more than 100 failures. This led scientists to try the simpler embryo-splitting method. In theory, the Dolly technique can produce entire herds of genetically identical animals, while embryo splitting has the potential to make only four clones at most.

However, even pairs of identical monkeys could be invaluable to medical researchers. For example, in AIDS vaccine research, immune system cells from a monkey given a test vaccine could be transferred to an identical monkey to see if those cells give protection from AIDS. Such an experiment could answer the question of what kind of immune system cell can protect against AIDS. Researchers could then turn their attentions to developing vaccines that stimulate those cells. ■

Genetically identical monkeys, two of the results of Oregon's cloning programme.

WHAT THEY DID

The Oregon researchers took young, eight-celled macaque monkey embryos and divided them into subembryos of two, three or four cells each. They transferred 13 of those partial embryos to the wombs of surrogate mother monkeys, and four of those became established pregnancies. Three of those pregnancies ended in miscarriage. The surviving cloned monkey is Tetra. ■

THE GENETICS NEWS ■ *Cloning news*

SPLITTING HEIRS

The shape of twins to come?

In theory, scientists could use the 'Tetra' technique to split human embryos to produce identical twins, triplets or quadruplets. At present this technique is not forbidden by legal restrictions on human cloning, although researchers at George Washington University caused controversy in 1993 by conducting human embryo splitting experiments.

However, the American Society for Reproductive Medicine (ASRM) has acknowledged the potential usefulness of embryo-splitting as a treatment for infertility, because a woman who produces a single embryo might be able to turn it into four.

Embryo splitting duplicates an embryo that has been conceived normally by two parents. It is not too different from the natural process by which identical twins are made. Dolly-style cloning, on the other hand, creates genetic duplicates of a single adult – an impossibility in nature. The creation of human clones by embryo splitting should therefore raise fewer ethical concerns than Dolly-style human cloning. ■

A DOTCOM START-UP

The birth on 5 March 2000 of Millie, Christa, Alexis, Carrel and Dotcom saw Blacksburg, Virginia, become 'Cloned-pig capital of the world'. Scientists at PPL Therapeutics, part of the Scottish Roslin Institute that cloned Dolly the sheep in 1997, announced last week that they had cloned five pigs at the Blacksburg research facility. The pigs were created by fusing an adult pig cell with another pig's egg cell, the DNA of which had been removed. The resulting embryo was then inserted in a surrogate sow.

In the past three years, sheep, cattle and mice have been cloned in Scotland, Virginia and Hawaii. Cloning pigs, however, opens up possibilities for cross-species transplantations because their organs are believed to be more compatible with human organs. The PPL team say their breakthrough could end a global shortage in hearts, kidneys, livers and other organs needed by thousands of people worldwide. Some estimates put the value of the pig-organ transplant market as high as £4 billion a year.

To breed successful pig organ donors, the scientists now have to genetically alter pig cells to make them more compatible with human cells. This includes eliminating a pig gene that adds a sugar group to pig cells that is rejected by the human immune system, and adding at least three genes to a cell that prevent delayed rejections of pig organs.

Dotcom and her sisters are revealed to the public.

Meanwhile, Dotcom and her siblings live a life of luxury in a secret barn in Blacksburg, hidden from animal-rights activists and other potential threats. Addressing the ethical issues involved in harvesting animal organs for humans, PPL point out that millions of pigs are bred each year for slaughter, while only 50,000 to 100,000 would be bred annually as organ donors. It would seem odd to object to a pig being used to save a life while you tuck into your eggs and bacon. ■

Health

MADE-TO-MEASURE MEDICINE

Genetics is showing pharmaceutical companies the way to make drugs that suit a person's genetic make-up.

Doctors have known for a long time that we don't all react to medicines in the same way. One of the reasons for this lies in our genes. When you go to the doctor in the future you might be asked to provide a sample for genetic testing before the doctor prescribes any drugs for you.

Time and money would be saved because drugs would not be tried out on people on whom they might not work at all or who might have a bad reaction to them because of genetic reasons. Within the next five to fifteen years the identification of genetic differences that are related to the diagnosis and treatment of disease will mean that clinical trials can be carried out more efficiently and drugs may reach the market faster as a result. Drugs will be marketed alongside tests to identify those individuals whose genes best suit them to the treatment being offered. The knowledge gained through genetics could be used to develop extremely effective drugs that can be prescribed to the largest possible group of people.

However, some people believe the drug companies may not want to move too fast down the road to personalised medicine. The risks are high and the drug companies don't want to face being sued if they get it wrong.

THE COST OF INEFFECTIVE MEDICINE

■ In the 1950s many patients died as a result of a reaction to a chemical used in surgery. They didn't have the gene that allowed them to clear the chemical from the body.

■ Around 100,000 Americans die each year because of reactions to drugs. This is the fourth biggest killer of people in the United States.

■ It is estimated that doctors in Britain waste £100m a year prescribing drugs that are ineffective or unsuitable for some patients. ■

Of mice and memory

A pill to restore the failing memories of elderly people may be a possibility following research carried out at University College London's Wolfson Institute for Biomedical Research.

Just like humans, mice begin to lose short-term memory and spatial awareness as they age. The Wolfson scientists found that switching off a gene involved in regulating levels of electrical activity in brain cells produced a mouse that keeps its short-term memory and other mental abilities into old age.

Dr Karl Giese said people had a similar gene, although it was not certain whether it had the same function. The Wolfson team is now trying to develop drugs that will have the same effect as the genetic change. ■

Individually tailored medicine within reach?

AN END TO INHERITED DISEASE?

A child with muscular dystrophy.

Towards the end of 1999 we came a step closer to being able to design the genetic make-up of our children. Scientists in Canada succeeded in breeding mice that had an artificial chromosome – a collection of genes assembled in the laboratory – inserted into their cells while they were still embryos. The new chromosome was reproduced in every cell of the fully-developed mouse's body. This technique, called germline intervention, would be illegal if carried out on humans. But will that always be the case?

Every human cell has 46 chromosomes. Scientists in Canada and the US are looking at the possibility of inserting a 47th chromosome that would be used to carry replacement genes into patients who might otherwise suffer from life-threatening genetic disorders. The hope is that artificial chromosome therapy would replace traditional gene therapy, a field beset by problems (see page 9). Gene therapy at present relies on viruses to deliver the therapeutic genes. Viruses are not ideal for two reasons. First, they can cause dangerous reactions in the host's body. Secondly, because they are very small, the number and size of genes they can deliver is limited. An artificial chromosome, on the other hand, can carry any number of genes and is not attacked by the body's immune system.

The artificial chromosomes given to the mice were transmitted from cell to cell during cell division and from one generation to the next as the mice reproduced. If humans were treated in the same way, not only would the embryo given the artificial chromosome be free of 'disease genes' its descendants would be too. Potentially, diseases caused by faulty genes, such as muscular dystrophy might be eradicated entirely. ■

Some genetic disorders that can be detected before birth

DISORDER	INCIDENCE
Cystic fibrosis	1 out of 2,500 white births
Down's syndrome	1 out of 800–1,000 births
Duchenne muscular dystrophy	1 out of 3,300 male births
Fragile X syndrome	1 out of 1,500 male births
	1 out of 2,500 female births
Haemophilia A	1 out of 8,500 male births
Huntington's disease	4–7 out of 100,000 births
Polycystic kidney disease	1 out of 3,000 births
Sickle-cell anaemia	1 out of 400–600 black births
Tay-Sachs disease	1 out of 3,600 Ashkenazi Jewish births

GM intelligence

Scientists at Princeton University have inserted extra copies of a gene into mice that appears to be linked with intelligence. The researchers tested the mice to assess various kinds of intellectual ability. The engineered mice were quicker to learn, and more likely to remember what they had learned, than 'ordinary' mice. All animals, including people, possess this gene. Will we soon be able to enhance our own intellect? ■

The BioSteel bucks

The Business Editor

It sounds like science fiction, but Nexia Biotechnologies of Montreal, Canada, are pinning their hopes for the future on a herd of African Dwarf goats that produce spider silk from their udders!

SPIDER SILK

The idea of creating a silk-producing goat started when Nexia researchers were working on genetically altered animals to produce blood-clotting agents for humans. At the same time, Dr Randy Lewis of the University of Wyoming was trying to interest people in his plans to produce spider silk outside of the spider. Spider silk is four times stronger than Kevlar, the material used for bulletproof vests, and it's light too. Nexia call it BioSteel.

Nexia's scientists took cells from a dwarf goat, inserted them into goat eggs and implanted them into a doe. They also threw in a gene from one of Lewis's spiders. Effectively what they had was a goat egg with 70,000 goat genes and a spider-silk gene. Five months later, Peter (named after Peter Parker, superhero Spider-man's secret identity) and Webster were born. Within a couple of years they had become the founding fathers

of a herd of transgenic goats producing spider silk protein in their milk that can be collected, purified and spun into incredibly strong fibres. The production from one goat in a month should be enough for an ultralight bullet-resistant vest. The silk could also be used for artificial tendons and ligaments.

LOSING THE BIOARGUMENT?

Bioindustrialists such as Monsanto have seen their stock prices collapse spectacularly as a result of the public backlash against GM foods. The protests have had such an impact that farmers in the US have reduced their GM plantings by 25 per cent.

To make matters worse, American farmers have filed a class action suit against Monsanto, charging them with inadequate testing and an attempt to monopolise the seed industry. The Iowa-based Alliance for Bio-Integrity has filed another

suit against the Food and Drug Administration accusing them of misrepresentation of risks and a violation of a law that requires demonstrated safety before marketing.

The biocompanies are driven to distraction by these claims. They say they've tested their products for years and that these accusations are based on fear and not on any evidence. But claims of potential dangers to monarch butterflies and to soil organisms from

Consumer pressure? Sainsbury's set up a helpline to discuss GM foods.

transgenic corn are hard to disprove without the benefit of time and money. Can the bioindustry afford them?

THE GENETICS NEWS ■ *Business and Farming*

Decision time

Farmers in the American mid-west have had to take a tough decision recently. Should they risk planting genetically modified crops, or not? Many are convinced that GM products give greater yields for lower costs. The problem is that like European and Asian consumers before them, the American public has become increasingly vocal in expressing their concern about these 'Frankenstein foods'. The autumn harvest could be plentiful but largely worthless.

CORN CONFUSION

The corn (maize) the farmers want to sow is known as Bt corn. It uses a gene derived from a soil bacterium, Bacillus thuringiensis, to make the whole corn plant toxic to the corn-borer, a caterpillar that drills into the corn cob and destroys the crop. The same bacterium is used as a pesticide by organic farmers. The protein that affects the corn-borer is the same in each case. One organic food store in Washington, DC, happily offers fruit and vegetables 'grown without pesticides' which, in fact, have been genetically modified not to need them. ■

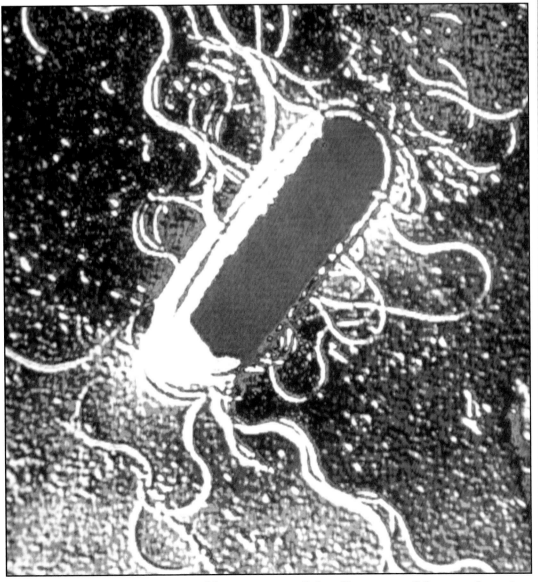

Bacillus thuringiensis – tiny soil bacterium and deadly enemy of the corn-borer!

THE PRICE OF PURITY

All crops, conventional or GM, cross-contaminate one another. Pollen can drift half a kilometre, so that one farmer's GM crop can cross-pollinate his neighbour's organic plants. Farmers selling non-GM foods fear that the value of their products will be lost if they are contaminated on the route to the consumer.

The only solution is to test for purity at every point in the supply chain, which may be too costly. GM products may in the end be abandoned not because they are unsafe but because the costs of keeping them apart from other crops are too high. ■

Soyabean success

American farmers are the world's largest producers of GM foodcrops. In 2003 nearly 60 million acres of American farmland had been planted with genetically engineered soyabeans, accounting for over 80 per cent of total soyabean production. ■

EU and GM

In July 2003 the EU passed laws requiring the labelling of all GM foods. The regulations mean that the food industry will have to keep GM and non-GM foods strictly separated. ■

The Culture

LIFE-LIKE ART, ART-LIKE LIFE?

Can genetics inform the creative world of art as well as science?

The Arts Editor

Genetic engineering is a highly specialised and expensive business, carried out in well equipped laboratories by trained scientists. For the moment some artists are making do with creating new life in the virtual reality world of computers. In its simplest form this might be seen in the fashion a few years ago for 'computer pets'. More serious work has been done in creating whole zoos of colourful artificial life forms that change and evolve along the same lines as organisms in our living world.

DESIGNER PETS

But is there a time coming when we might see sheep that glow in the dark and purple-spotted cats? Could artists and geneticists get together to produce 'designer pets'? The creation of new life forms has already occurred in the world of science. In the future, as the tools of genetics become more widely available, artists might be able to join in finding new ways of creating life too.

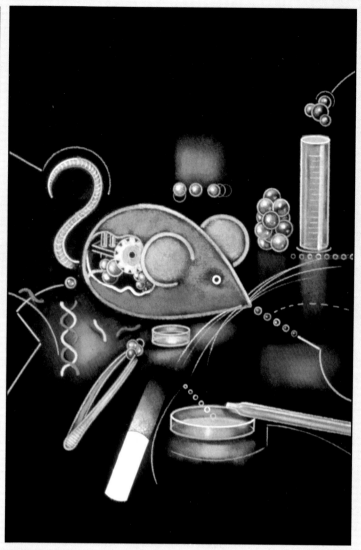

'Bionic Mouse' by Andrzej Dudzinski

GENE INSPIRATION

Here are four artists who have have been inspired by genetics.

Edward Steichen
In 1936 Edward Steichen displayed genetically-altered delphiniums at the Museum of Modern Art in New York. Steichen used traditional methods of selective breeding along with colchicine, a drug that altered the plant's genetic make-up to create delphinium flowers that had never been seen before.

Joe Davis
Artist Joe Davis works with genetic engineers at the Massachusetts Institute of Technology. He is, in a way, creating a new form of bacterial life by inserting a synthetic piece of DNA into an E. coli bacterium that will be reproduced and passed on to future generations of bacteria. He has coded a message into the DNA that can be read back and decoded as an icon he calls Microvenus.

Hunter O'Reilly
Hunter O'Reilly is a geneticist who is also an artist. The paintings she produces are inspired by watching cells grow under the microscope and by the structure of DNA. For Hunter, science and art complement each other. As she sees it, science is about finding order in nature and her art also 'invites the viewer to discover order, and to find order within themselves, where it is least expected'.

Andrzej Dudzinski
Polish born Andrzej Dudzinski is not a scientist and has not used genetic experimentation in his art. Rather he has used his paintings to explore some of the social and ethical issues raised by the new technology. Pictures like 'Bionic Mouse' (above) express both his wonder at the possibilities of genetic engineering but also his disquiet by emphasising its artificial aspect.

Genetic heroes

Science has often played a part in storytelling and movie making. To make a change, *The Genetics News* has asked our science editor to review some of the products of this sci-fi tradition.

Ethan Hawke stars in Gattaca as an 'in-valid'.

■ **Jurassic Park**, a novel by Michael Crichton made into a blockbuster movie by Steven Spielberg, did much to popularise the notion of genetic engineering with its plot involving the reconstruction of dinosaurs from prehistoric DNA. The genetic engineering is just a sophisticated plot device to get the dinosaurs in place. In previous years an intrepid explorer would have stumbled across a dinosaur 'Lost World' in the New Guinea jungle, perhaps.

Crichton, with degrees in physical anthropology and medicine, has the knowledge to get the science right. Except sometimes he doesn't, and he admits that he doesn't. At a meeting of the American Association for the Advancement of Science Crichton told scientists that movies are not supposed to portray science accurately. The problem is how is the audience supposed to know that? ■

■ **Gattaca** is a movie supposedly about the dangers of modern genetic research. Its characters inhabit a world of the not-so-far future in which the population has been divided into the physically perfect 'valids' conceived in test tubes and the less than perfect 'in-valids' conceived in the usual way. Beneath it all it's just another little guy taking on the big guys movie, with a murder mystery thrown in for good measure. In this case the little guy is, you guessed it, an in-valid and the big guys are represented by Gattaca, an all-powerful corporation staffed by valids. Guess who wins? As the movie subtitle puts it, 'There is no gene for the human spirit.' So that's all right then. ■

■ **Chromosome 6**, a novel by Robin Cook, features a race of transgenic apes created for organ transplants, but it's really a book about the Mafia being mean to everyone as usual. The genetics is just a sensationalist add-on. Nobel Prize-winning scientist James Watson has said, 'We used to think our future was in the stars, now we know it is in our genes.' Certainly genetics is playing as much a part in books and movies these days as space. However, the message in these adventures would seem to be that geneticists can't be trusted. Is this a true reflection of things or simply a cynical commercial pandering to public prejudices? And are these movies and books fuelling these prejudices still further? I would recommend the safest thing is to see them as no more or less than what they are – entertainment with absolutely nothing to tell us about science. ■

Jurassic Park: *the fact is, it's fiction.*

STONE AGE DNA

In Europe 30,000 years ago our ancestors mingled with another human-like species – the Neanderthals. For a long time palaeontologists suspected that the beetle-browed, stockily built Neanderthals must have interbred with modern humans. DNA from a 29,000-year-old fossil found in a cave in the northern Caucasus shows that it didn't happen.

Enough Neanderthal DNA has been recovered to allow Igor Ovchinnikov and his colleagues at the University of Glasgow in Scotland to compare key sequences with human DNA samples. The difference between human and Neanderthal DNA is enough to suggest that we have no Neanderthal coding in our genes.

The researchers estimate that humans' and Neanderthals' most recent common ancestor lived 150,000 to 350,000 years ago. However, some palaeontologists still dispute the team's findings.

Relax – this Neanderthal isn't one of our ancestors!

A GENETIC BIG BANG?

A bold new theory has attempted to tie together the origins of language, schizophrenia and the human species.

Professor Tim Crow, a research psychiatrist at Oxford University, has had a long-standing interest in schizophrenia and its causes. Schizophrenics make up about one per cent of any human population, a frequency that has remained more or less unaltered for centuries. Professor Crow has been developing ideas about its origins. The theory is that it is the trade off humans have had to make for the evolution of language. Professor Crow came to this idea by bringing together the fact that schizophrenia is a uniquely human condition and it is not found in our closest living relatives, the chimpanzees.

Schizophrenia appears to have a genetic basis. If one identical twin has the condition there is a roughly 40 per cent chance that the other twin will have it too.

All part of being human?

The problem is that schizophrenics tend to have fewer children than other members of the population, so why haven't the 'schizophrenic genes' disappeared over time? Professor Crow, thinks that language and mental illnesses like schizophrenia came about as the result of a single genetic change that happened about 150,000 years ago in a genetic equivalent of the big bang.

As a consequence of this mutation the left and right halves of the brain began to develop independently. The left half of the brain, where the language processing centre is found in most people, developed faster than the right half. The brains of schizophrenics tend to be much more symmetrical than those of other people.

Crow and his team tested thousands of British schoolchildren and found that those who were strongly right or left handed (which implies a strong left/right difference in their brains) scored highest on tests of verbal ability. Children who were close to being ambidextrous scored lowest on the tests. Significantly, schizophrenics are less strongly right-handed than the majority of the population and are often ambidextrous.

According to Crow, in schizophrenics the normal partnership between the right and left halves of the brain has been disrupted and it is this that results in hearing 'voices' and paranoia. The schizophrenic can't distinguish between his own ideas and those of others or between thoughts and speech. Crow says that this means that the brain's language controls have broken down.

Gardening

The summer sound of lawnmowers could become less of a feature of garden life following a discovery by a Washington University scientist. Michael Neff has discovered a gene that could cut down on the number of times gardeners need to cut the grass.

Neff, a plant molecular geneticist, found a gene that acts a bit like a volume control for plant growth. The gene works in tobacco and another laboratory plants. It has yet to be tested in grass but it might be only a few years before dwarf grass is made available to the public.

Plant growth is regulated by hormones, among them the brassinosteriod hormones, which stimulate cell growth. Neff and his colleagues focused on the growth of a mustard weed known as Arabidopsis, a plant that is commonly used in laboratories. They found a gene called BAS1 that instructs cells to produce a substance that destroys brassinosteroids. This reduces cell growth. They also found a mutant form of the gene that instructs cells in the leaves and stems to produce much more of the substance that destroys brassinosteroids.

DWARFING EFFECT

When they transferred the mutant gene into tobacco plants they found that some of the genetically engineered plants grew only 15cm or so tall rather than the normal 1.5 metres. This dwarfing effect was inherited by the modified plants' offspring. Neff is convinced that they should be able to do the same thing with horticultural and agricultural plants. Dwarf hedges might require less trimming. Fruit trees would be easier to harvest.

Neff aims to put the mutant gene into rice, a kind of grass. Then he will try it in poplar trees. He says, '… a lot of work needs to be done to address safety and marketability, as well as effects on disease resistance and other factors… the important thing is that we are beginning to develop a set of genetic tools that can be used to alter plant structure.'

Neff wasn't looking for a way to stunt plant growth. His aim was to learn more about how plants respond to light and that meant learning

about the growth hormones in plant cells and the genes that regulate those hormones. 'It was totally fortuitous,' Neff said. 'When you ask the big questions about how something works, you have no idea what the answer is going to be. But you know the answer is going to be exciting.' ■

A biogardener with GM tobacco plants – sowing genes takes the place of sowing seeds.

Gene Games

Will genetically-engineered super athletes win the race?

The Sports Editor

The Olympic Games of 2008 might just see an unprecedented number of records shattered. These could be the first games of the genetically-engineered super athlete.

UNDETECTABLE

Already the technology is emerging to produce athletes with bigger muscles and greater stamina than ever before. Unlike drug enhancements, genetic upgrades would be virtually undetectable. The proteins made by engineered genes are exactly the same as those made by the body naturally. The authorities will have a hard time keeping ahead of the game.

Gene therapies are currently being developed to fight genetic diseases, but why not use genetics to improve a healthy body and make it better? In the 1964 Winter Olympics Finnish skier Eero Mantyranta won two gold medals, due in part, perhaps, to a genetic advantage over his opponents. He was born with a mutant gene that resulted in his blood having 25 to 50 per cent more red blood cells than normal. Since red blood cells carry oxygen round the body it meant that his muscles got a better supply than they would have done normally, giving him substantially greater stamina. ■

Eero Mantyranta: a mutant Olympian hero.

ATHLETES BORN NOT MADE?

Scientists in Britain have discovered a gene that can increase muscle efficiency. The gene comes in two forms, a long one and a short one. Most people have one long and one short version. Those people lucky enough to have two long genes can build their muscles more efficiently with exercise than those with short versions of the gene.

However, it's not so bad if you have a double short gene. Researchers discovered that sprinters were most likely to have this combination. Long-distance runners, on the other hand, were more likely to have the double-long combination. ■

Born to run?
It's all in your genes.

A WAY TO WIN?

Biotech medicine is already finding its way into the changing room locker. Here are two examples:

As the competition gets tougher, will athletes be able to resist the temptation of undetectable biotech medicine?

Anaemia to athletics

In 1989 the biotech company Amgen began marketing an artificial form of the hormone that instructs the body to manufacture red blood cells. It was intended as a treatment for people with severe anaemia but athletes were quick to see the advantage it could give them. Athletes in track and field, tennis, skiing and football are all rumoured to use the drug and its genetic component means it is virtually impossible to detect. Soon, some suggest, gene therapy techniques could be used to make this type of advantage permanent.

Magic muscles

Researchers are working on a protein called MGF (mechano growth factor) to treat diseases such as muscular dystrophy. MGF is one of the proteins in the body responsible for repairing and building muscles after exercise. When scientists injected mice with the gene responsible for making MGF they found that their muscles grew by 20 per cent in two weeks. In theory athletes would be able to use MGF therapies to target specific muscles in their bodies, in effect growing muscles on demand.

WHO'S WHO

Here are some of the organisations you can go to for further information about genetics and the issues it raises.

■ **Friends of the Earth (FoE)**
Environmental pressure group who have taken an active role in the GM food debate. 26-28 Underwood Street, London N1 7JQ; TEL: 020 7490 1555
http://www.foe.co.uk/
FoE Australia 312 Smith Street, Collingwood, PO Box 222, Fitzroy 3065
http://www.foe.org.au/

■ **The Genetics Education Center of the University of Kansas Medical Center**
Links to a wide range of topics on genetics. It's aimed mainly at educators but students can find much of interest there too!
http://www.kumc.edu/gec/

■ **Center for Genetics and Society**
http://www.genetics-and-society.org
An organization that works to encourage responsible use of genetic technology.

■ **GM Nation? The public debate**
http://www.gmpublicdebate.org
The official website for the genetic modification public debate held in April 2003.

■ **The Human Genome Project**
For up-to-date information on the Human Genome Project visit its home page on
http://www.ornl.gov/TechResources/Human_Genome/

■ **The National Human Genome Research Institute**
Maintains a comprehensive list of links to genomic and genetic resources on the World Wide Web.
http://www.nhgri.nih.gov

■ **The Nuffield Council on Bioethics**
An independent body established by the Trustees of the Nuffield Foundation in 1991. Current members of the Council include clinicians, educators, lawyers, philosophers, scientists and theologians. Considers the issues raised by new developments in medicine and biology.
The Nuffield Council on Bioethics
28 Bedford Square,
London WC1B 3EG
TEL: 020 7 681 9625
http://www.nuffield.org/bioethics/index.html

AUSTRALIA

■ **Genetics Society of Australia Inc. (GSA)**
GSA aims to promote the interests of the Science of Genetics in Australia. Contact: Dr Geoff Clarke, PO Box 225, Dickson ACT 2062
http://www.gsa.angis.org.au

■ **Australian Gene Ethics Network**
A federation of groups and individuals in Australia who promote critical discussion and debate on the environmental, social and ethical impacts of genetic engineering technologies. 340 Gore Street, Fitzroy, VIC 3065 TEL: (03)9416 222
http://www.geneethics.org

■ **The Human Genetics Society of Australasia**
Provides a forum for those dedicated to the study, investigation and practice of Human Genetics. Also aims to promote public awareness. Contact: Ms Cherie McCune, Royal Australasian College of Physicians, 145 Macquarie Street, Sydney NSW 2000
TEL: (02)9256 5471
http://www.hgsa.com.au/

Note to parents and teachers
Every effort has been made by the Publishers to ensure that these websites are suitable for children; that they are of the highest educational value, and that they contain no inappropriate or offensive material. However, because of the nature of the Internet, it is impossible to guarantee that the contents of these sites will not be altered. We strongly advise that Internet access is supervised by a responsible adult. ■

YOUR VIEWS ON THE NEWS

The Genetics News doesn't just want to give you its views on the news. It wants you, its readers, to talk about the issues too. Here are some questions to get you going:

■ If you could be screened to find out if you were likely to suffer from a genetic illness would you do it?

■ Would you be prepared to pay extra for health care that was tailored to your genetic profile?

■ Do you see anything wrong with making salmon bigger? Would you eat a transgenic fish? And if not, why not?

■ What are the dangers of pollen from GM crops? Is cross-fertilization of one plant with another possible?

■ Should gene therapy experiments continue?

■ If people understood more about genes would they be less afraid of genetic modification?

■ What do you feel about breeding animals to provide life-saving drugs and transplant organs for humans?

■ If a foetus is tested and found to have a genetic defect is this grounds for an abortion?

■ Have movies and books influenced your views on genetics? If so, how?

■ What potential can you see for genetics in sport? Is its impact necessarily bad?

WHAT'S WHAT

Here's *The Genetics News'* quick reference aid explaining some terms you'll have come across in its pages.

■ **Alzheimer's disease** An illness of middle or old age in which the brain deteriorates and mental abilities are lost.
■ **amino acid** One of a group of naturally occurring chemicals used by living organisms to make proteins. Plants and some microorganisms can make amino acids but animals must get them from their food.
■ **bacterium** Any of a large group of single-celled organisms that do not have a distinct cell nucleus.
■ **bases** Parts of the DNA molecule that form the basis of the genetic code.
■ **biotechnology** Using biology for industrial purposes, especially in the genetic manipulation of organisms to produce medicines and other chemicals.
■ **blood-clotting agents** Substances present in the blood that trigger the formation of blood clots to seal a wound and prevent excessive bleeding.
■ **chromosome** Threadlike structure that becomes visible in the nucleus of a cell just before it divides. Chromosomes carry the genes that determine the organism's characteristics.
■ **clone** An organism produced from a single parent to which it is genetically identical.
■ **codon** A sequence of three bases in a DNA molecule that directs the placing of an amino acid in the making of a protein.
■ **conservation** Action taken to protect the natural world from the actions of humans, such as pollution and exploitation, that might threaten it.
■ **DNA** Deoxyribonucleic acid, a molecule found in almost all living things with the exception of some viruses, that carries the organism's genetic code, as a set of instructions for making proteins.
■ **egg cell** The female reproductive cell that is fertilized by a sperm cell and then divides to form an embryo.

■ **embryo** An organism in the early stages of development from the fertilization of the egg until it reaches a recognisable form.
■ **ethical** Of, or relating too, moral principles, the set of guidelines we follow regarding the behaviour that is approved of by society.
■ **European Union** An association of certain European countries formed to encourage free trade and close political ties between those countries.
■ **gene** A set of instructions for assembling a protein from amino acids. A gene is a length of DNA and a number of genes are carried on a chromosome.
■ **gene therapy** Treating an illness that is caused by a faulty or missing gene by repairing or replacing that gene.
■ **genetically modified (GM)** Describes an organism that has been changed by manipulating its genes.
■ **genome** The complete set of genetic material of an organism.
■ **germline** A series of germ cells, or sex cells (sperm and egg), each descended from earlier cells in the series.
■ **heredity** Passing on of characteristics from one generation to the next by way of genes.
■ **hormone** A substance produced in one part of the body and transported via the bloodstream to another part where it produces an effect, such as growth. Hormones are sometimes called chemical messengers.
■ **immune system** The body's defences against attack by microorganisms.
■ **infertile** Unable to reproduce.
■ **insurance premiums** An amount of money paid to an insurer, who will then pay an agreed amount of compensation in the event of damage or loss of property or health.
■ **molecule** The smallest unit of a compound or an element; a molecule consists of at least two atoms joined together.

■ **mutant** An organism that shows the effects of a mutation.
■ **mutation** A random change in the genetic material of an organism.
■ **National Health Service (NHS)** Government-funded health care authority set up in 1945 to provide free health care for all who need it.
■ **opiates** Drugs, such as heroin, that are derived from or related to opium in their chemical structure and effects.
■ **organic farmers** Farmers who produce crops or livestock without using chemicals, either as fertilisers or pesticides or for any other purpose.
■ **patent** A licence granted by the government to an individual or organisation giving them the right to make and use an invention.
■ **pesticide** A chemical designed to kill insects or other pests that damage crops or livestock.
■ **pharmaceutical** Of or relating to pharmacy, the act of preparing and dispensing drugs for medical purposes.
■ **pollen** The plant equivalent of sperm, produced by the male part of a flower.
■ **protein** Any of a number of very large molecules that perform vital tasks in living things, including regulating the speed of chemical reactions and providing part of the organism's structure.
■ **receptors** Organs or cells in an organism that respond to outside stimulation and then send signals along the nerves to trigger a reaction elsewhere in the organism.
■ **staple crops** Crops of vital importance as a source of food.
■ **surrogate** A substitute or stand-in for something or someone else.
■ **synthetic** Describing something similar to a natural product but which has been made by combining two or more chemicals.
■ **transgenic** Describing an organism that carries genes from more than one species.

INDEX

ageing 20
AIDS 18
Alliance for Bio-Integrity, Iowa 22
American Society for Reproductive Medicine 19
art 24
athletics 28-9
Avery, Oswald 15

Baker, Bruce 17
Bateson, William 15
Bee Farmers Association of the UK 7
bees 7
BIO 2000 10
Biodevastation 2000 10
Blackburn, Elizabeth 17
books 25, 30
Boveri, Theodor 15
Boyer, Herbert 15

Caplan, Arthur 16
Caskey, Charles 15
Celera Genomics 14, 15
Chromosone 6 25
cinema 25, 30
Clinton, President Bill 14
cloning 3, 9, 11, 18-19, 31
Cohen, Cynthia 17
Cohen, Stanley 15
Collins, Francis 14
Cook, Robin 25
Council on Responsible Genetics, Cambridge, Massachusetts 9
counselling 5
Crichton, Michael 25
Crick, Francis 12, 15
Crow, Tim 26

'designer babies' 17, 21
'designer pets' 24
DNA 4, 8, 9, 12-13, 14-15, 19, 26, 31
Dolly the sheep 3, 18-19
Dotcom the pig 3, 18-19
drug addiction 6

embryo-splitting 18-19

English Nature 6
ethics 4, 16-17, 19, 31
European Commission 11

famine 10
fish, genetically modified 6, 30
Food and Drug Administration (USA) 22
food, genetically modified 10, 11, 16, 17, 20, 21, 22-3, 27
Friends of the Earth 7, 30

Gattaca 25
Gelsinger, Jesse 9
gene therapy 9, 15, 20, 21, 31
'germline' gene therapy 17, 21
Giese, Dr Karl 20
goats 22

Hawtin, Geoffrey 10
heredity 8, 12-13, 15, 16, 21
honey 7
Hubbard, Ruth 9
human genome 3, 10, 12, 14-15, 30
Human Genome Project, the 14, 15, 30
Hunt, Stephen 6

illness 4, 9, 15, 16, 21, 30
Institute for Genomic Research, Maryland, USA 16
Institute for Public Policy Research 5
Institute for Social Ecology 10
Institute of Forest Protection and Management Research, Korea 9
International Plant Genetic Resources Institute (IPGRI) 10

Japan 23

Jurassic Park 25

Kahn, Alex 11
Korea 9

Lander, Eric 11
Lenaghan, Jo 5
Lewis, Dr Randy 22
Lysenko, Trofim 8

MacLeod, Colin 15
Massachusetts Institute of Technology (MIT) 11
McCarthy, Maclyn 15
medicine 3, 6, 9, 15, 20
Mendel, Gregor 8, 15
mice 20, 21
Miescher, Johann 15
monkeys 18
Monsanto 22

National Health Service 4, 5, 31
Native Americans 8
Nature 6
Neanderthals 26
Neff, Michael 27
Nexia Biotechnologies, Montreal 22

organ transplants 19

pigs 19
pollen 7, 23, 30, 31
Public Health Genetics Unit 4

Radar 5
Russia 8

schizophrenia 26
screening, genetic 4, 5
silk 22
spiders 22
Sutton, Walter 15

Tokar, Brian 10
trees, genetically modified 9, 27
Tuva 8

University College, London 6
University of Pennsylvania 9, 16
University of Washington 27
University of Wyoming 22
US National Bioethics Advisory Commission 11
US National Human Genome Research Institute 14

Vavilov Institute of General Genetics, Moscow 8
Venter, Craig 14

Watson, James 12, 15, 25
Wellcome Trust, the 14
Wilmut, Dr Ian 17
Wolfson Institute for Biomedical Research, University of London 20

Zakharov, Ilya 8
Zimmern, Dr Ron 4